Critical Ethics for Engineers in 2026 and Beyond

Critical Ethics for Engineers in 2026 and Beyond

Keith Warwick, P. E.

This edition has been published by Central West Publishing PTY LTD, (ABN 13 683 898 722) Australia
© 2026 Central West Publishing PTY LTD

All rights reserved. No part of this volume may be reproduced, copied, stored, or transmitted, in any form or by any means, electronic, photocopying, recording, or otherwise. Permission requests for reuse can be sent to info@centralwestpublishing.com

For more information about the books published by Central West Publishing PTY LTD, please visit https://centralwestpublishing.com

Disclaimer
Every effort has been made by the publisher, editors and authors while preparing this book, however, no warranties are made regarding the accuracy and completeness of the content. The publisher, editors and authors disclaim without any limitation all warranties as well as any implied warranties about sales, along with fitness of the content for a particular purpose. Citation of any website and other information sources does not mean any endorsement from the publisher, editors and authors. For ascertaining the suitability of the contents contained herein for a particular lab or commercial use, consultation with the subject expert is needed. In addition, while using the information and methods contained herein, the practitioners and researchers need to be mindful for their own safety, along with the safety of others, including the professional parties and premises for whom they have professional responsibility. To the fullest extent of law, the publisher, editors and authors are not liable in all circumstances (special, incidental, and consequential) for any injury and/or damage to persons and property, along with any potential loss of profit and other commercial damages due to the use of any methods, products, guidelines, procedures contained in the material herein.

A catalogue record for this book is available from the National Library of Australia

ISBN (print): 978-1-922617-78-1

The author maintains no liability for information presented in this book. The technical information has been presented solely for teaching purposes and is not intended to be used for any construction, planning, design, engineering, or similar activity.

Preface

Critical Ethics for Engineers in 2026 and Beyond is a frank and candid description of the need for an increased level of ethics in the engineering and construction industry. It describes how to become an ethical person rather than just following rules established by others. It explores the depths of life which is where an ethical commitment is formed. It empowers engineers to formulate ethical standards and solutions to address societal changes posed by global warming and war. It elucidates the contribution of values, morality, and religion towards maintaining a high level of respect for engineers.

The title actively supports the tenets of Diversity, Equity and Inclusion and focuses on the treatment of women in the engineering and construction industry. The advocates of women's rights in the sixties and seventies did greatly improve the living and working conditions in the home and workplace, but posited that women are the same as men and to be successful a woman had to live as one, which is untrue and unfair. Women should be given a few days off each month with pay to cope with menstrual distress. They should not have to ask permission, obtain a signature from their boss, or even explain why they will be gone. To avoid embarrassment, awkwardness, and bias they should simply be able to leave a yellow stickie on their desk or a note on their calendar that they will return when they are able to. They should not be expected to check-in with the office, return calls, write emails, or review reports.

The book elucidates the contribution of values, morality, religion, and conscience towards developing an ethical platform. It includes detailed discussions of vital subjects such as ethical character formation; protection of the public and workers; mentoring and being mentored; integrity; sustainability and ecology; colleagues, leadership team, and clients; transparency; environmental justice; ability to perform work; and forward-looking case studies.

Table of Contents

1	Ethical Character Formation	1
2	Advocating for the Client	7
3	Mentoring	13
4	Integrity	17
5	Sustainability and Ecology	23
6	Relationships	29
7	Transparency	33
8	Environmental Justice	45
9	Functional Capacity Evaluations	53
10	Scenarios	57

Chapter 1

Ethical Character Formation

Engineers desire to be ethical in their personal and professional lives, but it is a determined process as opposed to a natural state. Fortunately, there are many sources of guidance on how to become an ethical person.

Codes of ethics such as those written by the American Society of Civil Engineers in 1914 and the National Society of Professional Engineers in 1964, are industry standards. Collectively, these codes of ethics address a myriad of topics common to the four major branches of engineering: civil/structural, mechanical, electrical/ computer and chemical as well as the dozens of focused specialties.

Several other groups such as the Institute of Electrical and Electronics Engineers (IEEE); American Society of Mechanical Engineers (ASME); and American Institute of Chemical Engineers (AIChE) have developed their own codes of ethics which reflect the challenges encountered within that sector of engineering. These guidance documents provide direction on how to avoid harmful situations that they could find themselves embroiled in.

These professional societies research and develop technical standards; write protocols; lobby for passage of laws and regulations that will benefit society; and host seminars and workshops. These gatherings provide opportunities to share experiences, discuss new technologies, join committees, collaborate on books and articles, and develop long-term professional relationships. The presentations on successful projects reflect the need for ethical conduct among the many parties involved in a successful venture.

Many professional organizations publish magazines such as Civil Engineering, PE Magazine, IEEE Spectrum, Mechanical Engineering; and Chemical Engineering Progress which contain articles on specific aspects of ethics within their sector of the engineering profession. These publications, as well as many others, offer engineers the opportunity to write articles and journal entries on a subject of their choice. This enables engineers to become a more skilled and more ethical engineer. Many will research a subject, write an article, and interact with those who benefit from your knowledge.

Many of the leading certification organizations such as Leadership in Energy and Environmental Design (LEED), which is sponsored by the U.S. Green Building Council, require an organization to be ethical as well as technically competent. Becoming aware of and implementing their principles into your professional life will mold your character.

Self-study of books on ethics is an effective way to learn because you can do so in comfortable surroundings, be alone, and read at your own pace. This independent approach allows you to go on tangents into nuances which pique your curiosity. There are quite a few books on engineering ethics, some quite old, and others that have been written in the past few years. Some are generalized in nature and others are written for a specific profession or to remedy a certain condition. Some stay close to the surface which is quite safe and non - controversial; and others delve into the philosophy of ethical behavior in engineering and explore controversial topics which can lead to conflict. Many of the new books delve into the uses of and limitations of artificial intelligence (AI) which is a topic of concern.

For generations established engineers have been mentoring young men and young women who have stepped into the professional arena. An older ethical engineer will pick a young engineer or two and guide them through the early years of their career. He will help them intertwine their personal, academic, and professional lives to achieve an effective work/ life balance and practice self-care. They will teach them how to design a sewage treatment plant (wastewater treatment plant) but will not do the calculations, but instead will allow him or her to learn how to do so. They will describe their failures, many of which were caused by a breach of ethics. This wisdom cannot be absorbed from a book but can only be gained by interacting one-on-one with a mentor. We may only remember a structure that we designed for a couple of years after it has been built, but we will think about lessons learned from a mentor throughout our career.

There are many dedicated classes on engineering ethics taught at every level of higher education so if you are still in college, working on a PhD at night, been terminated, watched your company go out of business, or have taken a couple years off of work to further your formal education you can easily take those classes. The textbooks that you purchase will become some of your most valuable resources as you advance in your career. Ideally you will take classes from professors who are licensed engineers or have had a significant amount of professional engineering experience.

There are many excellent continuing education classes on engineering ethics. Some are offered for free or as part of a subscription service. On - demand courses (recorded) tend to be much less expensive, but if you learn more effectively from a live person it may be worth the fee. If your supervisor is aware of the importance of ethics, he will probably pay for it. Ethics within the engineering profession is so important that many licensing boards require that you take an ethics class to become registered and each time you renew your license.

One way to become ethical is to observe how ethical professionals conduct their affairs. Energy engineers at the iconic Empire State Building located in midtown Manhattan, which was built in one year and forty-five days starting on March 17, 1930, conducted a thorough energy upgrade of the massive structure in 2010 which included installation of an elevator heat recapture system, use of reflective barriers on radiators to reduce heat loss, an adaptive dimming system which adjusts in response to the level of natural lighting, energy efficient lighting, and retrofitting the 6,514 windows in the building. Considering that energy reduction ultimately reduces the need to burn fossil fuels, this was an ethical undertaking which we can all be inspired by.

In 1962 Rachel Carson, a biologist, courageously wrote Silent Spring which described how DDT almost caused birds such as the dynamic Bald Eagle (whose numbers had dropped to under 2000 individuals in the lower 48 states in the early 1960s) and the magnificent yet fragile Californian Condor (whose numbers had dropped to about 20 individuals in the early 1980s) to become extinct. DDT caused the eggshells to thin allowing the chicks to break out early and die from exposure to harsh environmental conditions. She was criticized by those who used DDT but was supported by President Kennedy. Her efforts energized an intense grassroots environmental movement that lasted until the mid 1970s. Her need to tell the public about the dangers associated with DDT, which had been heavily used since the 1940s, led to the formation of the Federal Environmental Protection Agency in 1970, passage of the Clean Air Act in 1970, enactment of the Clean Water Act in 1972, and implementation of the Endangered Species Act in 1973.

Our nation, since its inception on July 4, 1776, has been an ethical one. We have a moral heritage that is intertwined with mainstream American culture. The United States Constitution signed September 17, 1787, and many other key governmental documents are platforms for ethical behavior. The desire to be so in our career and personal life reflects our society's tradition. Performing ethical actions such as advocating for environmental justice, which is the premise that the poor, minorities, marginalized, and disenfranchised, deserve as clean an environment as the middle-class, upper-middle class, and the wealthy. An extension of this premise is that these disadvantaged people should have enough money to live comfortably. Practicing eco - justice, or any other worthwhile endeavor, contributes to the development of an ethical character.

It is important to be transparent. It is advisable to let our co-workers and managers know what projects we are working on, how much of it we have completed, and any concerns we have. When they understand that we have a problem they can approach us with a solution. Some people are shy and need to be prompted to give and receive advice. Most long-time engineers have developed a sense of what is considered ethical within the engineering profession. Your supervisors have probably found themselves in the same quagmire that you are in such as having to coordinate with a city official who refuses to accept a letter of explanation or realizing you may have made a design error. They know how to anticipate a problem, guide you away from a compromising situation, maintain your credibility, avoid a precarious condition, and keep you from violating your basic integrity.

Academics and researchers who study and write about ethical matters related to engineering are a valuable source of guidance. The field of ethics, which is an outgrowth of morality, requires us to recognize the inherent value of mankind including all races, nationalities and genders, the standards of professional practice, and procedures on how to make practical decisions. Ethics includes several branches and frameworks including: metaethics which is the study of morality; normative which is the study of appropriate actions; applied ethics which is an intense focus on a specific subject such as engineering; utilitarianism which addresses the satisfaction of those impacted by our actions; deontology which discusses meeting others people's needs; and virtue ethics which describes the character of the person making an ethical decision.

Because we have a conscience most people inherently know when their actions have been or will be less than optimal. The conscience, which complements morality, is an internal indicator of the acceptability of our viewpoints and philosophies. It is not a totally reliable indicator of whether we are living in an ethical manner, but it does serve the purpose of prompting us to utilize additional methodologies to determine the suitability of our choices.

It is advisable to stay within the bounds of your qualifications when selecting projects because when undertaking a project that is too difficult it can become very hard to maintain an accurate perspective. You may find yourself taking unethical actions because you may not know enough about the project to make sound decisions. It is not wrong to limit your design activity to small expansions of existing buildings. The stress of designing a $70,000,000 luxury hotel can be enormous and only the most qualified engineers should undertake such a challenge. Subjecting yourself to that level of stress will likely lead you to break your own ethical codes and make some very unwise decisions. We can easily learn to be ethical from designing several small projects rather than from one large effort.

Becoming ethical can be an avant-garde process of exploring new ways of finding solutions. We want to sail far above the mundane, shallow, and tedious activities of the day and explore the depths of life which is where an ethical commitment is formed. While you may have an epiphany which helps you to commit to leading an ethical life it usually does not happen as such. It requires proceeding on a long and sometimes steep path towards excellence in your job or business.

Engrossing yourself in books on the philosophies of engineering ethics, theories of ethical behavior, types of ethics, history of ethics, business ethics, case studies of successful projects, ethics in construction, construction contracts, and professional service contracts will provide you a framework in which to make the difficult decisions that you will inevitably have to make during the course of your career.

Chapter 2

Advocating for the Client

Engineers are responsible to advocate for our client, to look out for his best interests, and help him to avoid failure of any kind. Our responsibility extends well beyond preparing clear plans (blueprints) and specifications.

We endeavor to protect him from making technical mistakes and from hiring a disreputable contractor, dishonest superintendent, or incompetent subcontractor. We must try to prevent him from entering into a venture that he does not have the wherewithal to succeed in and from starting a project that he cannot afford to complete.

While every engineer relishes a new project including the gross revenues, expansion of his portfolio and reputation, a fresh new computer file, and a blank piece of bond paper, vellum, or mylar; it is unethical to begin the design process and bill him accordingly when what he is asking for is not realistic and the project will fail miserably. The scope of the effort may be too large and too expensive to ever become profitable or to allow him to recoup his initial investment. His entire idea, dream, or plan may not sync with reality or align with current societal patterns, trends, and movements.

If he is naive about construction, engineering, business, applicable laws, or life itself, and is spearheading the project by himself, you may suggest that he get help. You may suggest that he find a partner; use a construction manager - at - risk project delivery method; or hire a construction consultant, construction attorney, or an architect, to walk him through the process. He may not realize how tough the construction industry and the manufacturing industry are, and while most are honest, there are some that are not. The client must understand the fast pace of work and while people in construction do get long breaks between projects, when on a job they work very hard.

The engineer needs to prepare plans and specifications that are so clear and unambiguous that the contractors will bid lower because the risks presented by uncertainty become minimal. The engineer should label all dimensions to keep the contractor's cost estimator from having to scale distances which can lead to imprecise measurements or catastrophic mistakes. He needs to make sure all disciplines of engineers communicate with each other before and during the design process. It can extend the critical path several days when there is a lack of collaboration.

If the client tasks an engineer with work that is out of his area of expertise and qualification, he needs to advise his client to hire another engineer even if he loses thousands of dollars of revenue.

You should design machines or computer hardware or buildings that can be built easily and quickly. Unless the client suggests irregularly shaped walls or curved surfaces, it is best to standardize the design to keep the contractor from getting confused, making mistakes, or submitting the dreaded change order. Ensure that unique items, such as an antique component, can be obtained within a reasonable period of time before complying with the client's wishes and specifying one. The plans and specifications should highlight this unique and complex interior feature so the contractor can bid and schedule delivery in a prudent manner.

He may not understand that obtaining a building permit can take months or even a year if the jurisdictional authority is busy or short staffed and the plans and specifications need to be returned multiple times for corrections.

At times we need to be candid and point out ongoing conflicts between individuals and organizations that are involved in the project. You might know about these but your client has no way of detecting them. The building official may have been against the project for many years and is simply not going to cooperate. You should develop strategies to bypass and overcome the obstacles so that he can avoid temporary or permanent delays.

Artificial intelligence is of little or no value to engineers. Engineering design software is often AI-assisted so the engineer will lose control of the output. The engineer needs to check any results generated by AI or design software. In addition to a check of the math, algebra, geometry, and trigonometry the engineer needs to refine his estimate by use of hard copy cost estimating guides, personal knowledge of the project cost, knowledge of similar products, comparisons to similar estimates in the firm's database, and consultations with senior engineers.

There are significant drawbacks to using AI to write content for reports, studies, or letters. The output may flow nicely and contain few, if any, typos or grammatical mistakes, but the text will not be project specific and will not include nuances or personal touches that only a human writer can add. The client will likely be able to tell that the content was created by AI software rather than a human by using AI detection software program(s) and noting that the style of the report is inconsistent with standard engineering tones and wording. This will undermine your credibility and professional standing.

An ethical engineer will avoid displacing people from a house their family has owned for generations or is the only place they can afford to live. He will try to prevent businesses or government officials from demolishing the structure to build something larger and more luxurious. The owner may not want to sell the house which can prevent a project from moving forward or leave a small house pathetically placed adjacent to a factory which is problematic for all. That conundrum can reduce the chances of completing the building and reduce profits in the outyears. The project owner will lose clients; suffer a loss of credibility, and incur the wrath of the community.

The National Historic Preservation Act of 1966 gave birth to the National Register of Historic Places and local extensions of that legislation. That Act has protected many historic structures but there are many that have been missed. An engineer is responsible to protect those even if the law does not do so which can result in a change of project scope and lowered fees. A collection of regulations has been passed to protect the ambience of these historic neighborhoods which would prevent him from building a modernistic glass and steel building amidst century old craftsman homes.

Until the 1950s land developers and realtors often signed restrictive covenants that prevented certain ethnic and racial groups from purchasing property in white neighborhoods. This practice led to the formation of racially segregated neighborhoods, many of which remain in place today. An engineer should advise his client not to build in a neighborhood that illegally discriminates against any prospective residents.

While the traditional and long-standing design - bid - build project delivery method is the most common, there are other delivery methods that have come into vogue. One is the construction manager - at - risk method which is advantageous for those who are uncomfortable with the construction or machine prototype design process, even though it can lead to higher project costs and less control of them by the owner. To proceed down this path the owner must hire a construction manager in the planning phase and allow him or her to shepherd him through the planning, engineering, construction, and warranty phases of the construction effort. Another is the design - build concept which allows one company to provide both the engineering and construction services to the owner thus eliminating the potentially conflictual bidding process. A drawback of the design - build model is that the construction company is responsible to hire the engineer which makes it harder for the engineer to maintain his independence. At times he will be prompted to design structures that will be more profitable for the contractor as opposed to being in the best interests of the project owner. When working within a construction manager - at - risk, or design - build model the engineer must endeavor to maintain his professionalism.

The time and material contracting method is used for unique computer/electrical and mechanical design processes. It is incumbent upon an engineer when contracted to design prototype components under a time and materials contract to record work hours and tasks accurately and only spend time performing work that is delineated in the scope of work. It is unethical to assign a person to a project full time and bill the client accordingly, and then ask them to perform other work at certain times during the day.

The contractor is allowed to consider documents provided to him by the owner to be correct, but that is often not the case, so the engineer should determine which ones can be relied upon and which ones should be redone. Until the late 1960s and early 1970s, when pocket calculators were released slide rules were used almost exclusively by engineers to perform calculations. The level of slide rule accuracy varied in accordance with the engineer's skill, care, and eyesight. Older engineers certainly used one early in their career and often display it in the lobby of their office as an interesting piece of history. Calculations and plans (blueprints) prepared using slide rules should be redone if a high level of accuracy is needed.

Unless a land survey has been completed recently, the engineer should suggest that the owner have one done. Older surveys may have been performed using manual transits and surveying rods which could be inaccurate if the rodman had to hold the rod underwater, in a wet area, or on a high spot or low spot on the ground. The measurements were taken by stretching a metal tape from one point to another and reading the measurement by dropping a plumbob on the corner. The measurement could be inaccurate if the surveyor could not pull on the tape with sufficient force to prevent it from sagging. If the rodman inadvertently held the rod in the wrong location the distances measured would contain error.

It is intensely frustrating for a project owner to deplete the project funds before finishing the building. The engineer is responsible to alert him of that eventuality so that he can make financial maneuvers to save the project. If he is operating as a sole practitioner, as opposed to a corporation, his personal finances, if he has any, could be in jeopardy. The sooner he begins to take mitigative actions the greater his chances of making a profit or at least avoiding bankruptcy are. In any case, he has bleak prospects to face and difficult decisions to make. One option, with the approval of the lender, is to de-scope the project such as building seven houses as opposed to the ten he planned. The lender wants him to succeed so they may be open to extending the terms of the loan, giving him extra time to make payments, or suspending payments until he can generate revenue from the building or invention.

The engineer should advise him not to borrow money from friends or family because if there are changes in relationships or family structure he might have to return the money prematurely. He may need to hire a construction consultant to pull him out of the quagmire. He could sell the unfinished building but will lose any profit or overhead he had hoped to receive and may have wasted months or years of his life and career.

Chapter 3

Mentoring

We all want a legacy. We want people to remember us fondly. Our legacy is what we have done for others including our family, friends, colleagues, and clients.

Anything we publish, teach, design, or say becomes part of the corporate memory and body of knowledge of the engineering profession as a whole and available to the millions of engineers alive at any one time. A more focused and manageable group, and one that is easier to define, is the dozens of the people in our firm, firms we have worked for in the past, partners, clients, and consultants that we know personally. Our mentoring efforts will take place within that subset of the profession or ecosystem.

Engineers with experience have been pressured to alter designs, falsify numbers for political or economic reasons, exaggerate, tell white lies, and accept a bribe. They have faced many ethical dilemmas, handling most of them quite well, but at times have found themselves embroiled in several types of quagmires, which they had overcome. This seasoning, survival, and prospering establishes their credibility and enables a senior engineer to mentor those who are in the early stages of their career.

The author worked hour - by - hour for a few months with a seasoned construction superintendent who had lived a century in 20 years. The author was the lead field engineer on the project. During this mentoring relationship he taught me to be confident, work safely, not be a victim, to not be afraid to tell an equipment operator not to drink beer onsite, and how to treat people with respect. He told me about his failures and struggles and how he had overcome them. Each day began with breakfast at a diner near the site and ended with an ice cream cone on the way home. Not everyone gets to work so closely with a mentor but for those who do it is an unforgettable experience. It has been over 40 years but I still remember him so very fondly.

Mentoring must be performed on a one - to - one basis. Interacting with two or more young engineers at the same time can be instructional, but it is not an effective form of mentoring. Mentoring involves tailoring what is said to the temperament and needs of the young engineer which is difficult to do when interacting with more than one person at the same time. The mentor can share specific personal experiences that will benefit the young man or woman which is very difficult to do within a group. An older engineer can mentor more than one person at a time, but it must be at separate times in different locations. Sometimes an older engineer does not have the time or energy to mentor more than one person so it may be best to focus on just one relationship.

Sometimes a young engineer will approach an older one and ask for guidance, or a member of the leadership team may contact the young man or woman and offer to assist. At times a relationship can develop naturally by the junior engineer just gravitating towards a man or woman in mid-career. The mentoring relationship may be a result of a random assignment of both to the same project or it could be systematically planned. The mentor will not get promoted or compensated for his efforts, but it will be one of the most rewarding aspects of a 40 year career. The mentoring process can take place by phone, email, or zoom but it is much more effective when it is accomplished in person.

The relationship may begin with the mentor teaching the junior engineer practical aspects of the engineering process such as that artificial intelligence has limited value within the engineering profession and the results calculated by engineering software need to be checked. It can include teaching the entry level employee how to write a concise report for a member of the leadership team and how to write a brief, concise, and clear memo. The new graduate may have specific project related questions which the mentor can help him with or tell him where to find the answer. He will help him understand how to approach a project, collaborate with peers, and deliver a finished product that has been checked, polished, and written in an easy-to-understand style. There may be times when the young engineer is frustrated and needs some encouragement which the mentor can give him.

A mentor should be candid about his protege's abilities and determine what he is particularly skilled at and direct him into careers that he will enjoy and be successful at. He may suggest projects that are simpler than most, smaller in scope, and quite short. Some engineers are more suited for tasks that are largely non - technical such as marketing, finance, safety management, environmental compliance, teaching, and writing. A good mentor will assess his personality and direct him towards work that will complement rather than conflict with his personality. He will help his protege to intertwine his natural abilities, skill levels, career goals, and personal life in such a manner that he will succeed and prosper throughout his career.

Oftentimes the best advice someone can give you is that you are not suited for the position that you are in. It is trite and misleading to say that anyone can do anything if they try hard enough. It will be unpleasant to tell a young man or young woman that they are not skilled enough or do not have the personality traits necessary to perform satisfactorily in the department that they are in. Even so, it must be done to save them from many years of unhappiness.

While mentoring must be performed on a one - to - one basis it can take many forms. When working in the same department it may consist of allowing the new graduate to observe how his mentor manages different sets of circumstances. It can consist of a more formalized arrangement where the mentor meets once or twice a week to enjoy a snack and talk about how he is doing. Or it may just be performed on an as - needed basis where the young engineer will approach the mentor when he has a personal or professional problem or concern. The process can take place within a few months or be extended to a decade or more.

Over time the young employee may open up to the seasoned veteran about his personal life which will give the mentor the chance to teach him about work/life balance and self - care. There is great satisfaction in watching a young person learn from you and adopt some of your personal and professional practices which have been honed for decades. A mentor should not tell a young engineer that he has been mentored if it was accomplished in tandem with multiple people. To do so would prevent him from seeking an effective mentoring relationship and cause him to use the wrong approach when it is his time to be a mentor.

A mentor should teach his young student the importance of honesty and integrity and not to enter into conflict unless it is absolutely necessary, but in any case, to remain firm and present a strong defense. He will help him to increase his level of emotional intelligence and to stay calm even when faced with the most toxic co-workers. He will make sure that he has a good work ethic and does not spend his time on unimportant tasks.

The mentor can, if given the opportunity, advise him on how to form effective relationships with his family and friends. He may be able to advise him on how to gather a few partners and form a consulting firm. Some young adults will prosper if they own their own business or perform freelance work, although there are risks associated with leaving an established company.

Chapter 4

Integrity

An engineering company owner told the author that "we sell integrity" An engineer needs to perform his own functional capability evaluations to ensure he or she is able to perform his or her responsibilities. He should voluntarily withdraw from a project or company altogether if he or she becomes unable to perform his assigned tasks. This could be due to driving restrictions; physical inability to walk through congested factories, construction sites, and laboratories; mental or physical illness; severe dizziness; and uncorrectable impaired eyesight.

An engineer should design each facility in a green and sustainable manner that will facilitate its certification as a green building by the U.S. Green Building Council, Leadership in Energy and Environmental Design (LEED) organization or other established certification agency. LEED certification is required by many cities such as New York, Miami, San Francisco, and Atlanta for certain projects before they will issue a building permit. LEED offers several levels of certification including platinum (most stringent), gold (second most stringent) silver (third most stringent) and certified (least stringent). To be certified a building must demonstrate compliance with the following precepts: Locating buildings near public transportation systems, designing sustainable and green buildings, implementing water conservation, establishing energy conservation protocols, protecting indoor air quality, and addressing environmental concerns.

It is incumbent upon engineers to carry errors and omissions insurance with a limit sufficient to pay for the cost of an error that he could reasonably be expected to make on a particular project. It is also prudent for an engineer to carry workman's compensation insurance. Cybersecurity insurance is important to protect confidential employee and client information. Some organizations require an engineer to carry sexual misconduct and molestation insurance. The author feels that no one should offer engineering services to an organization that requires it because it can be supportive of a child molester.

Plans (blueprints) and specifications should be checked by at least one professional engineer who has a sufficient level of expertise in the subject. It is not acceptable to ask an unlicensed engineer to do so.

It is essential for an engineer to be honest, not exaggerate, and not withhold important information. He needs to be forthright in all matters of conduct and in all situations. He needs to pay his bills on time as opposed to waiting until he gets paid by a client. He must properly classify a worker as an employee or consultant and manage each one accordingly. He should only take credit for bona fide experience.

His fees should be relatively consistent with other local engineers that have the same qualifications, and he should not bill unlicensed and clerical workers at his rate.

He should avoid the use of Artificial Intelligence (AI), fully check the output of engineering software, and have another licensed engineer check his manual calculations. AI is not advisable for design processes because it is often incorrect and uses unnecessarily long calculation methods which are difficult to verify. Plans and specifications should be prepared in an unambiguous manner to prevent inflated bids, confusion, delays, lawsuits, mediation, arbitration, and litigation.

When an engineer realizes that he has made a design mistake he is obligated to report it to the project owner immediately to prevent a prototype from being developed in an improper manner.

Many municipalities have procedures and requirements that are more stringent than those of the government or those found in industry standards. It is prudent to learn about these by reading the procedures and requirements and speaking with the applicable authority prior to beginning the design process.

He should not make handwritten notes on plans that will be used to prepare bids or for construction purposes. They can be hard to see and read which can result in bids that are much higher than expected due to the uncertainty.

When an engineer discovers a mistake that a worker has made, he should inform him or her prior to contacting their superior or describing it in an inspection report. In most situations the worker can promptly eliminate the problem or establish a recovery plan that is acceptable to his supervisor.

An engineer should make a concerted effort to locate underground utilities before beginning the design process. He should look for disturbed or sunken ground, signs, markings, manholes, cleanouts, and connections to buildings. He is advised to ask the owner, past owner, architects, contractors, environmental inspectors, building inspectors, utility companies, realtors, city officials, neighbors, and safety inspectors about buried pipes on the site. He should review plans, specifications, geological reports, environmental reports, notices of violations, land surveys, municipal utility locating services results, construction as-built drawings, utility easement documents, documents prepared by utility companies and past utility companies, property records, tracer wires, digital utility maps, geographic information system databases, legal descriptions, photographs of pipe interiors, observe above-ground connections to utilities, determine what the utility needs of adjacent facilities are, listening devices, results from utility locating services, building condition reports, testing results, shop drawings, contamination surveys, electromagnetic locators, ground penetrating radar results, thermal imaging devices readouts, acoustic systems results, and building permits prior to beginning the design process.

The engineer should review the documents prior to designing the facility to ensure the information is relevant, current, correct, verifiable, approved, prepared by a qualified person, legible, and complete.

Pipes to search for include fire protection water, drinking water, slurry, chemical lines, storm water, irrigation water, natural gas, electrical, petroleum, communication, gasoline, diesel fuel, sanitary sewer, and steam.

It is essential to establish a file system for the project as soon as the client contacts you. This should be backed up by a secondary or tertiary electronics system and paper files. This should include specifications; land surveys; plans; as-builts: contracts; shop drawings; geological reports; environmental reports; budget reports; calculations; software outputs; notes; injury and illness reports; notices of violation; lawsuits; records of phone conversations; photos; screenshots; messages; videos; logs; texts; sign in sheets; training records; faxes; emails; tax forms; bank account numbers; permits; correspondence with the client, general (prime) contractor(s), and subcontractor(s); and communications from local, state, and federal officials. Folders should be opened for attorneys; medical professionals; the media; engineers; architects; suppliers; manufacturers; buyers; investors; lenders; insurance companies; bond companies; and results of building, environmental, and safety inspections.

The reasons for keeping an extensive and organized set of each revision of each document is to ensure that construction is carried out in accordance with the plans and specifications; warranty issues can be handled fairly; lawsuits, arbitrations, meditation, and the full litigation process can be resolved in a legal and fair manner; accurate as-builts are prepared, field decisions can be made quickly and accurately; corrections to the drawings can be made promptly; calculations can be re-checked; and the engineer and attorney can respond quickly and accurately to any freedom of information act request, and notice of violation.

When selecting the format of the project files the engineer should remain aware of the limitations of electronic storage methods. Hard drives and other computer components typically fail within a relatively short period of time. These limitations are caused by wear and tear, drops, bumps, spills of liquid, high temperature, high humidity, dust, power outages, misuse, and power fluctuations.

Cloud services, which include extensive data storage capabilities, do experience periodic and temporary outages. These are from human error, misconfigurations, hardware failures, server failures, and inadequate software.

Buildings that store hard copy and electric media are subject to failure by floods, water leaks, wind, hurricanes, tornadoes, earthquakes, ocean actions, geothermal activity, rain, snow, lightning strikes, heavy live loads, fires, geological failures, wear and tear over a long period of time, poor construction, civil disobedience, purposeful damage or destruction, and contamination so measures should be put in place to protect files from one or more of these failure modes.

It is important to obtain a P.E. license in one or more states. Some pass the exam without studying but many need to do so and there are some engineers who will never pass the fundamentals of engineering and professional engineer exams. There are many self-study programs, private tutors, and classes to help a young engineer prepare for these. Becoming a P.E. will allow you to stamp your own plans and specifications; establish credibility; be able to bill at a much higher rate; open a consulting firm with other engineers, and work as a subcontractor.

There are many excellent continuing education classes available on various engineering subjects. Some are offered for free or as part of a subscription service. On - demand courses (recorded) tend to be much less expensive and more convenient but the material in them can be outdated. If you learn more effectively from a live instructor or facilitator it may be worth the fee which your firm may pay. These one-hour or less on-demand and live engineering classes and webinars are taught by many public and private organizations and are ideal to watch at lunch and the last hour of the workday.

A safety factor (a 100% safety factor is double) should be assigned to all designs. The engineer's selection of the safety factor should be solely based on safety concerns and not driven by economic factors. He should consider the consequences of failure such as human injury or illness, damage to the environment and wildlife, and interruption of an eco-system.

Chapter 5

Sustainability and Ecology

An engineer should avoid accessing an environmental restoration project location, construction site, or industrial facility that is unsafe and should ask the contractor to remediate the unsafe areas. Those include conditions that violate OSHA 29 CFR 1910 (general industry standards) such as inadequate number, placement, and type of fire extinguishers; roof openings or floor openings not barricaded; single boards used for climbing; stairs without handrails; exit routes blocked and not passable; evacuation and exit route directional signs missing; hazardous materials stored in an unsafe manner; leaking drums of hazardous waste; machine guards missing; personal protective equipment (PPE) in poor condition; walkways blocked by equipment, material, and electrical cords; inadequate electrical safety; poor sanitation; and inadequate fire protection. An environmental restoration project location or construction site must have adequate programs in place to address 29 CFR 1926, Construction Safety, requirements including scaffolds; fall protection; excavation and trenching safety; welding and cutting safety; proper storage and use of compressed gas cylinders; use of motor vehicles and construction equipment; use of power, powder-actuated, and hand tools; shaded break areas with chairs and tables; clean and fresh water adjacent to workers; first aid kits; safety and environmental compliance postings; safety data sheets in work areas; and spill kits.

An engineer should ensure that no one must use a porta-potty that does not flush because standard models are unsanitary, insulting, and demoralizing.

There are many laws that prohibit an employer from requiring an employee to speak English at work, but many of them contain exceptions. These include requiring workers on environmental restoration project locations and construction sites to speak native English without an accent because employees need to be able to alert co-workers of impending danger such as a piece of earthmoving equipment moving towards them. An employee needs to speak quickly and clearly without delay to prevent an accident or death.

We want to treat everyone fairly and not ask anyone to deny their nationality and its traditions, but it is in everyone's best interest for employees to reduce their accent while at work. This will improve their confidence and help them to avoid the annoyance of people asking them to repeat themselves. It is difficult for teenagers and adults to reduce their accent so their sincere efforts and skilled coaching will not fully remove it, but the techniques used by voice coaches will help minimize it. It is prudent for the owner of an environmental restoration firm or construction company to hire individuals to assist his employees to speak in a clear manner.

An engineer should seek to learn about the cultural backgrounds of colleagues and offer to take part in or support their celebrations. These include holidays such as Cinco de Mayo, Mexico's Independence Day, Japan's Shogatsu, Chinese New Year, Yom Kippur, Rosh Hashanah, Kwanzaa, Midsummer, Norway's Constitution Day, New Years Day, Easter, July 4th, Thanksgiving, Christmas, Bastille Day, St. Patrick's Day, Ascension Day, Good Friday, Valentines Day, Father's Day, Mother's day, Summer Solstice, and Winter Solstice. In addition, co-workers enjoy celebrating birthdays, professional milestones, anniversaries, retirement, winning an election, birth of a child, graduation, promotions, marriages, awards, earning a black belt in martial arts, and winning a softball tournament.

Sustainability includes environmental and social fairness and appreciating other cultural groups character traits, traditions, and life skills. These include respect for the elderly, the value of hard work, the importance of education, family, natural healing to supplement traditional medicine, etiquette, the importance of rest, the advantages of peace, respect for leaders, harmony, humility, strong personal relationships, flexibility, warmth, hospitality, resilience, openness to others, optimism, resourcefulness, frugality, passion, the love of art and music, and creativity. It is important to understand how much of American culture is derived from other people's anthropological and ethnic traditions and values.

A key ecological concept is providing workers with safe and healthy work areas. Offices and common areas should be furnished with non-plastic chairs, tables, shelves, cabinets, flooring, and wall coverings that do not offgas hazardous vapors such as volatile organic compounds. VOCs are an irritant which can impair neurological functions, damage the liver and kidneys, and cause cancer. Instead of plastic try using untreated wood, metal, wool dyed with non-hazardous dye; cotton dyed with non-hazardous dye; and untreated rattan. Only non - hazardous cleaning supplies should be used in the building. These include plant-based wipes, non-toxic and biodegradable cleaning solutions, organic butters and oils, and organic lemon juice. The janitorial staff should avoid using chlorine and ammonia. Sometimes just soap and water are sufficient to clean dirty areas and surfaces. Because lye is used in soap making there is no truly organic soap, but it can be made with organic material.

Compassion is part of sustainability. We salute our older friends, mothers, fathers, grandparents, and great grandparents for surviving the agricultural depression and poverty among minorities and blue-collar workers in the 1920s, the depression from 1929 to 1941, WWII which took place from 1941 to 1945, and the Korean War which lasted from 1950 to 1953. Men, in particular, were affected by the economic hardships and wars which understandably made them less apt to open up to others and form mutually beneficial relationships at work, and more likely to strictly direct their employees without really relating to them or knowing them. Men chose to not show emotion and not be particularly nice which could be seen as a sign of weakness. The youngest WWII veterans are 95 and have left the workforce but not so long ago. The current company owners were trained by their fathers to adopt a rigid lifestyle, philosophy, and persona. However, some have moved past that firm leadership style to one that involves listening to their employees and considering their personal needs. Our newer corporate leaders are willing to promote an effective work/life balance and self-care.

To increase morale in the workplace, each employee, including receptionists, should have their own, single-occupant office with solid walls, a door, room for an extra comfortable chair, and printer(s). They should be encouraged to decorate their office with collectible furniture, decorative items, art, plants, and family photographs.

Employees who walk, run, or bike to work; and walk at lunch reduce their carbon footprint, but become quite grungy in the process. It will be harder for them to feel fresh, crisp, alert, creative, and clean throughout the remainder of the day. Many people will avoid them if they are unkempt or smell bad. To counteract that an employer should install a shower for employees.

It is necessary for an engineer to be familiar with Federal environmental legislation and the corresponding state and local regulations that must be as stringent or more stringent than the Federal government's. He also must be aware of international environmental protocols and agreements.

An example is the National Historic Preservation Act of 1966: This established a partnership between the Federal government and state/local agencies that were tasked with preserving historic structures and neighborhoods. The EPA charter was to review any proposed Federal action that involved earth moving or construction, which is the case in almost any proposed project. They became responsible for validating the scope of the project before money would be spent preparing the plans (blueprints) and construction specifications. A historic structure or neighborhood is one that is already on the National Register of Historic Places or meets the requirements to be added to it. An NHPA 106 review is required to be part of any level of National Environmental Policy Act of 1970 (NEPA) evaluation: Categorical exclusion which is for an action that has no significant effect on the environment; Environmental Assessment which is required for actions that could potentially have an adverse effect on the environment; and Environmental Impact Statement which must be performed for Federal actions that will have a significant negative impact on the environment. When there is going to be significant environmental damage the scope of the project must change, the design of the structure must be modified; or the project may be canceled.

A fictional example is that no one can erect a building designed in a modernistic style within a neighborhood of 120-year-old homes. If the tallest buildings in the neighborhood are three stories high, then the proposed building cannot exceed that height. A building or neighborhood sets a tone. It can be an aggressive corporate enterprise, or a peaceful residential neighborhood known for its gentility. These settings cannot be intermingled but must remain physically separated.

Rachel Carson, *Silent Spring*, 1962: This prominent book described how DDT almost caused birds such as the Bald Eagle (whose numbers had dropped to under 2,000 in the lower 48 states in the early 1960s) and the magnificent yet fragile California Condor (whose population had dropped to about 20 individuals in the early 1980s) to become extinct. DDT caused the eggshells to thin allowing the chicks to break out early and die from exposure to harsh environmental conditions. She was criticized by those who manufactured and used DDT but was supported by President Kennedy. Her efforts energized an intense grassroots environmental movement that lasted until the mid 1970s. Her need to tell the public about the dangers associated with DDT, which had been heavily used since the 1940s, contributed to the formation of the Federal Environmental Protection Agency in 1970, enactment of the Clean Water Act in 1972, and implementation of the Endangered Species Act in 1973.

While there has always been a concern for the environment there were three periods of focused attention from 1900 to the present. *The first movement* started in the early 1900s while Theodore Roosevelt was running the country and John Muir was leading the Sierra Club which he formed in 1892. It was a wilderness conservation and preservation effort which remained intense until WW1 began in 1914. *The second movement*, which was designed to protect soil, air, water, and wildlife, began when Rachel Carson wrote *Silent Spring* in 1962, and lasted until the mid-1970s. *The third movement,* which began about 20 years ago, is focused on minimizing global warming and is still a vital effort.

We are obligated to protect our ecology and ecosystems; live in a green manner; and comply with the Sustainable Development Goals (SDG) developed by the United Nations. These include no poverty; zero hunger; good health and well-being; quality education; gender and racial equality; clean water and sanitation; affordable clean energy; desirable job and business opportunities; economic growth; sustainable cities and communities; responsible consumption; and climate protection.

Chapter 6

Relationships

I will outline our obligations to treat our colleagues, leadership team, and clients with kindness and compassion. An ethical engineer will always try to maintain a sense of reasonableness when interacting with others. This includes using reasonable tones, body language, and wording in all manner of communications. He may need to decline a project which he feels is unethical such as designing a prison, but he should do so in a respectful manner and provide an alternative scope of work. I plan to list strategies on how to cope with mean and toxic people.

Women should be given a few days off each month with pay to cope with menstrual distress. They should not have to ask permission, obtain a signature from their supervisor or even explain why they will be gone. To avoid embarrassment, awkwardness, and bias they should simply be able to leave a yellow stickie on their desk or a note on their calendar saying that they will return when they are able to. They should not be expected to check-in with the office, return calls, write emails, or review reports.

There is a myriad of relationships that we must navigate and circumvent. Following are key people we will interact with on a regular basis

Accountants	<ul><li>Compassion</li><li>Privacy</li><li>Working conditions</li><li>Visibility</li><li>Recognition</li><li>Creativity</li></ul>
Managers	<ul><li>Respect</li><li>Transparency</li><li>Obedience</li><li>Few if any details</li></ul>

Construction workers	• Safety • Comfortable working conditions • Cold fresh water • Not be exploited • Representation • Reasonable expectations • Appreciation • Business opportunities
Architects	• Creativity • Privacy • Close companions • Compassion • Comfortable working conditions • Business opportunities
Engineers	• Compassion • Privacy • Creativity • Close companions • Comfortable working conditions • Personal development • Opportunity to get published • Opportunity to teach
Environmental scientists	• Compassion • Status • Privacy • Opportunity for advancement • Safety • Security • Comfortable working conditions • Close companions
inspectors	• Not be intimidated • Relied upon • Recognized • Low - stress
Construction managers	• Quality • Updates

	• Respect • Order • Cooperation
Project owners	• Initiate • Respect • Ideas • Assurance • Low - stress • Liked

Chapter 7

Transparency

It is imperative that engineers act in a transparent manner during their interactions with their partners, employees, consultants, suppliers, general contractors, construction sub-contractors, leadership team, clients, regulators, the media, the public, neighbors, local businesses, contracting officers, activist groups, union leaders, and governmental officials. Following are steps that we can take to be transparent.

Table 1. Openness, Honesty, Scrutiny and Apologies

Step	Discussion
Be open and visible	Make a concerted effort to share information in a timely manner
Be honest	Do not ever tell a lie no matter how insignificant it seems. Do not withhold information because that can be considered being untruthful
Be open to scrutiny	Allow people to question your judgement and disagree if warranted
Be willing to apologize if you are actually in the wrong	"I am sorry that you feel that way" is not an apology but is simply shifting blame. Start with "I am sorry that I" Apologies bolster a relationship

Table 2. Brainstorming, Contact Information, Observations, Sneakiness, Approach, and Data

Step Discussion

Have frequent brainstorming sessions	Allow engineering managers and staff to hear each other's unfiltered ideas concurrently
Share your personal phone number with text capability, home address, and personal email address with key personnel	Be available during off hours to hear their concerns and progressive ideas that cannot wait until Monday
Allow people to observe you work and be open about what you are doing	Invite them to a conference or a high-level meeting with government regulators and clients to learn about how you communicate with different audiences
Do not be sneaky	That is the epitome of a lack of transparency
Be approachable and welcoming to those who need to speak with you	Some may need some time in your office and some matters can be resolved in less than a minute during a discussion in the hallway
Provide raw data to those who need it	Many people want to see testing data, log sheets, and results of chemical analysis in addition to summaries and conclusions

Table 3. Critical Information, Safety Factors, Memos, Downturns, Personal Thoughts and Admitting Mistakes

Step Discussion

Step	Discussion
Do not withhold information from someone because you feel the person will not understand it; will react badly; and it will hurt you financially	We should give people the benefit of the doubt and know they can hire a consultant if they do not understand the documentation. We are obligated to provide information that people have a right to see even if it will impact your finances
Tell people what the safety factor for a structure is	This is a key piece of data that is needed to perform a risk analysis
Minimize the number of times that you write a private interoffice memo	Interoffice memos should be distributed broadly
Tell people as soon as you learn that a facility or business may shut down	Inform your staff about plans to close part or all of a facility and business
Share your thoughts, dreams, concerns, and visions	Sharing the more personal aspects of your personal and professional life will endear people to you
Admit your mistakes	Seek advice so you can minimize the number and intensity of mistakes that you make. When you do err let people know and tell them what you are doing to undo the damage. People are willing to forgive honest errors but not so willing to forget when someone hides them

Table 4. Nondisclosure Agreements, Decision Making, Invoicing, Body Language

Step Discussion

Step	Discussion
Avoid initiating or signing nondisclosure agreements	These prevent you from communicating openly and expose you to litigation if you inadvertently violate one. These are designed to limit liability and negative publicity as opposed to furthering honest and transparent communication
Describe your decision-making process clearly rather than being vague and ambiguous	This will help people understand what is important to you and know what is expected of them.
Tell a person they made a mistake before you include it in a report or tell their manager	People will appreciate the professional courtesy and the opportunity to resolve the matter before their supervisor hears about it
Send invoices out on time and do not hide and compile charges	Let people know what your billing is each month
Try not to be intimidating	Appropriate tone of voice, body position, and facial expressions can relieve an employee's anxiety

Table 5. Perfection, Informal Communication, Reserved Colleagues, Simple Documents, Change of Mind

Step Discussion

Step	Discussion
Do not try to be perfect or be too nice	It can be intimidating and annoying. Employees have difficulty relating to someone who tries to do everything right and is too nice. Avoid gushing or offering too many compliments

	within the same conversation
Linger at the water cooler and make light conversation, on Monday ask how their weekend was, attend office and personal social events, and invite people to important events such as weddings and anniversaries	It is easier for someone to discuss formal office matters if they have learned to be comfortable speaking with you
Encourage quiet people to talk more	It may be easier to talk with them on a one - to - one basis. Try asking simple and non - intrusive questions to begin the conversation
Do not cloud or complicate documents with unnecessary information	The reader may not need technical depth, background, history, your opinion, and personal examples
Tell people when you have changed your mind and why you did so	If your new decision affects the employee help them to avoid any negative consequences

Table 6. Avoid Bias, Decision Making, Change in Key Staff, Company Problems

Step Discussion

Do not be biased	If you are biased people will see that you are and it will undermine your credibility. It will affect your decision making ability. Bigotry can undermine the cohesiveness of an organization
Involve people in decision making processes	People enjoy being asked for their opinion and will readily share their candid thoughts.

Critical Ethics for Engineers

	People will be less apt to feel blindsided by a decision if they have been part of the decision process. Employees usually know more about their operations than managers so their input will be invaluable
Tell your staff if the president or owner is having serious health problems, planning to retire, or going to get replaced	Changes in key personnel can have a drastic effect on an employee's tenure and morale
Inform employees if someone is cited by a regulator, arrested, fined, had their license suspended, having financial problems, or will be involved in a scandal.	Employees will find out about these concerns eventually and it builds trust if you inform them early

Table 7. Schedule, Product Lines, Interviews, Staffing

Step Discussion

Provide people timelines of certain upcoming events	It is not helpful to say an event will occur after a certain date because people need to know how long after. People need to know what the earliest and latest dates an action will take place are so they can feel informed and not be frustrated
Tell employees about proposed new product lines and services	This will enable your staff to prepare for their new positions This expanded company portfolio may create opportunities for advancement

Tell employees that you are interviewing potential workers and introduce them on their first day	Assure your staff that the new employees are not going to replace them if that is indeed the case
Inform your staff if your company is considering reorganizing, merging, downsizing, moving, or closing offices	This allows people the maximum amount of time to reorder their personal and professional lives

Table 8. Firing, Truthfulness, Personal Life, Vision

Step Discussion

Tell people when they need to find a different job and help them to do so. Let them know if they are going to be fired or laid off and help them to manage this potentially disastrous scenario	The early notice will give them more time to rearrange their life. When people see you warn other employees of their demise they know they will not be blindsided
Do not tell different employees different versions of the truth and do not talk about people behind their back	Doing so is gossip which is inappropriate for a leader
Tell people what is important to you and place family photos, awards, and other personal objects in your office	Employees want to know what you do after 5:00 pm. People will respond to a person that they know personally
Share your victories and let people see you enjoy your career	People want to be happy for you when you achieve a level of success and believe that they will as well

Table 9. Objectivity, Victories

Step Discussion

Perform engineering work in an objective and professional manner and do not allow yourself to succumb to political or financial pressures	Engineers are expected to design safe and environmentally sound facilities and buildings and cannot be prevented from doing so by those who are politically ambitious and overly concerned with becoming wealthy These overly ambitious people will ask you to design buildings that will not be able to withstand strong winds, seismic activity, live loads, dead loads and the weight of the building as required in the design basis. They will ask you to compromise your integrity and put others at risk for profit.
Share your victories and let people see you enjoy your career	People want to be happy for you when you achieve a level of success and believe that they will as well

Table 10. Discretion, Options, Complexity, Intelligence

Step Discussion

Do not act on behalf of a person without their permission because it can be considered a breach of confidence	Quoting a person to a colleague on a sensitive or risky matter can severely impact that professional relationship
Let people know what the options that you are considering	People need to communicate their concerns to an

are when confronted with a corporate challenge. These include mergers, closing, selling, entering bankruptcy, changing product lines, downsizing, and a large and rapid hiring campaign	engineering manager regarding the consequences of a proposed action
Do not hide behind complexity and do not be too complicated. Allow yourself to be simple occasionally which is a quality that people are drawn to.	Being complex makes it harder for people to get to know you. Do not confuse people with too much data. Explain why you are sending out a communication which will satisfy their curiosity
Do not hide behind your intelligence. Feeling that you know more than everyone else can lead to diminished morale	Confusing colleagues with your knowledge will prevent them from approaching you because they feel they cannot compete

Table 11. Confidence, Questions, Risks, Specifics

Step Discussion

Tell people when you are sure a project will be successful and when you are skeptical and are only participating on a trial basis	This allows employees to perform their work and make plans based on the uncertainty
At the end of a conversation or meeting ask if there are any questions	This allows you to dispel any misconceptions
Be candid about risks and describe how a failure would affect you, the company, and each employee	Some people cannot emotionally, financially, and physically cope with a downturn and may need to seek a more secure work environment
Use specifics such as we built 14-	People usually like to know

inch diameter storm sewers at the intersection of Maple Street and Elm Lane; Main Street and Jones Alley; and First Avenue and Greenville Road as opposed to saying we built storm sewers at three locations	how much a project will cost and how long it will take to complete

Table 12. Notes, AI, Email

Step Discussion

A personalized hard copy typed letter warrants much more attention than an email or text. A handwritten note reflects your desire to develop a closer professional relationship with the recipient	People are more apt to respond to and act based on personalized hard copy letters and handwritten notes
Avoid using AI	It cannot include company specific information and does not allow you to communicate in a creative manner. People can spot it and will feel like you did not want to take the time that would be necessary to communicate with them in a personal manner
Personalize email messages	It is still useful as long as it does not have a spam-like quality. Do not send emails that are standard text with just their name typed in, often in a different style and font. Allow purple to communicate with you in their preferred format

	rather than be constrained to enter text boxes on a company platform

Table 13. Notice, Honesty, Personal Interests, Objectivity

Step	Discussion
Tell people when they need to find a different job and help them to do so. Let them know if they are going to be fired or laid off and help them to manage this disastrous scenario	The early notice will give them more time to rearrange their life. When people see you warn other employees of their demise, they know they will not be blindsided
Do not tell different employees different versions of the truth	Doing so is gossip which is inappropriate for a leader
Tell people what is important to you and place family photos, awards, and other personal objects in your office and share your victories	Employees want to know what you do at home
Perform engineering work in an objective and professional manner and do not allow yourself to succumb to political or financial pressures. They will ask you to compromise your integrity and put others at risk for profit.	Engineers are expected to design safe and environmentally sound facilities and buildings and cannot be prevented from doing so by those who are politically ambitious and overly concerned with becoming wealthy. These overly ambitious people will ask you to design buildings that will not be able to withstand strong winds, seismic activity, live loads, dead loads and the weight of the building as required in the design basis.

Chapter 8

Environmental Justice

For millennia the poor and minority communities have suffered injury and illness due to bigotry, slavery which continued into the mid 1900s, white privilege, and caste systems.

As a component of the civil rights ecosystem, environmental justice, which is the premise that members of the poor and minority communities should have as clean an environment as the middle class, upper middle class, and the upper class, became a visible cause in the 1960s. Until that time the health of members of impoverished communities was damaged because they were forced to breathe air containing fumes from municipal waste and hazardous waste incinerators and drink water contaminated with lead, arsenic, mercury, microorganisms, pesticides, and drug residue. The 1960s was a terrible decade filled with intense disrespect for the government; drug use; and indolence; but it did give rise to a few honorable causes such as environmental justice. This movement was initially energized by the radical infrastructure of the decade but lost focus as our society became hyper materialist in the late 1970s and 1980s.

The tenets of the environmental justice premise were born from an intermingling of civil rights and environmental protection and were finally recognized, validated, and formalized by President Bill Clinton when he signed Executive Order 12898, Federal Actions to Address Environmental Justice in Minority Populations and Low-Income Populations on February 11, 1994. President Joe Biden signed Executive Order 14096 "Revitalizing our Nation's Commitment to Environmental Justice for all" in April 2023 to validate and ensure compliance with the requirements set forth in EO 12898.

Synopsis of EO 12898:

REQUIREMENT	DISCUSSION
Section 1: Implementation	Each Federal agency shall achieve Environmental Justice and identify and address conditions that negatively affect the health of minorities and the poor in the United States and its territories and possessions, Washington DC, the Commonwealth of Puerto Rico, and the Commonwealth of the Mariana Islands Within three months of Executive Order 12898 the EPA will establish an EJ working group comprised of representatives from several U.S. agencies. Following that each agency is required to develop an EJ strategy
Section 2: Federal Agency Responsibility for Federal Programs	Each agency shall seek to obtain EJ within their agency.
Section 3: Research, Data Collection, and Analysis	Each agency shall conduct and analyze EJ related research, data collection, epidemiological studies, and clinical studies
Section 4: Subsistence Consumption of Fish and Wildlife	Consumption patterns can affect the health of various species and ultimately endanger human health due to buildup of contaminants in animals and fish
Section 5: Public Participation	Engineers should be transparent and share raw scientific data and prepare objective summaries that could be easily understood by the general public
Section 6: General Provisions	EO 12250, "Leadership and coordination of nondiscrimination laws" must

Critical Ethics for Engineers

	remain in effect EO 12876, "Historically black colleges" must remain in effect

Implementation of Environmental Justice by Engineers:

ACTION	DISCUSSION
Design, construct, operate, maintain, and demolish facilities in a manner that protects the health of poor and minority residents	We need to advocate for the health and safety of poor and minority populations by contacting lead agencies and lobbying politicians
Design, construct, operate, maintain, and demolish facilities in a manner that protects the safety of poor and minority residents	In particular this applies to industrial facilities that are adjacent to or downwind from low-income and minority neighborhoods
Design, construct, operate, and maintain drinking water treatment plants and secondary containment basins in a manner that provides clean drinking water for low-income and minority residents	Professional engineers shall design the same type of facilities necessary to protect water quality for wealthy Americans, poor Americans, and residents of third-world countries
Engineers shall operate, monitor, and maintain, air testing devices in a manner that protects the air quality of poor and minority residents	Professional engineers shall design the same type of facilities necessary to protect air quality for wealthy Americans, poor Americans, and residents of third-world countries
Engineers should design, construct, operate, and maintain facilities intended to treat, store, and dispose of hazardous waste in a manner that protects human health and safety, and minimizes the risk of chemical releases that	Engineers shall control the hazards presented by the hundreds of hazardous chemicals such as hydrofluoric acid (it is so dangerous it should be banned), mercury, lead, other heavy metals, lead, forever chemicals, and other acids.

would impact low-income and minority residents	
Engineers should advocate for reductions in the number of permitted commercial hazardous waste incinerators	There are about 20 permitted hazardous waste incinerators in the U.S. Although the number is less than the amount in use in the 1990s and 2000s there are several that are increasing capacity and will remain in operation indefinitely.
Professional engineers should specify the use of non-hazardous cleaning supplies in various buildings and facilities	This practice should be included in the general conditions of contracts and in maintenance manuals
Engineers should specify the use of furniture and wall coverings that are not made of material that off-gases hazardous vapors	Engineers should encourage architects, interior designers, interior decorators, and owners not to use those materials

EJ needs a little boost which engineers can provide. We should prevent hazardous waste landfills, and any type of landfill for that matter, from being built in or near minority and low-income neighborhoods. Over time many or most landfills do leak which will contaminate soil, surface water, and groundwater used in homes which will cause significant health problems. Our efforts should also lean towards removing lead and other contaminants from individual homes and public areas in minority and low-income neighborhoods; and preventing trucks loaded with thousands of gallons of hazard liquid from being driven through those neighborhoods.

The following information is courtesy of the Environmental Justice Foundation

Column 1 contains paraphrased information from the Environmental Justice Foundation, London, England, https://ejfoundation.org	Information in Column 2 was written by the author
The global ocean absorbs about 25% of the CO_2 generated from human activity	CO_2 causes acidification which is the process of ocean water becoming more acidic. Acidification affects organisms like oysters that make hard shells by combining calcium and carbonate from seawater. As acidification increases carbonate ions bond with excess hydrogen which results in less carbonate ions available to build shells. Engineers need to design structures and develop processes that significantly reduce the generation of carbon dioxide because the ocean is not a finite resource. Engineers need to consider carbon dioxide generation during all phases of the construction cycle including planning, design, procurement, construction, operations, demolition, and disposal of debris. An estimated 70% of the buildings in the U.S. are single family homes with an average life span of between 125 and 175 years so the debris that will be generated from demolition or destruction by natural disaster is significant.
There is a declining	Energy engineering is key to protecting

ocean fish population due to overfishing, climate change, and habitat destruction. A significant percentage of the people on Earth eat fish as a major source of protein.	the oceans and marine life. Energy managers reduce the amount of power required for industrial operations thus reducing the amount of carbon dioxide generated by burning fossil fuels. It is necessary to develop industrial processes that require less power and generate less CO_2. We need to defend international wilderness areas from deforestation which releases CO_2 and reduces the number of trees available to absorb CO_2. Burning and disposal of agricultural waste needs to be managed to minimize CO_2 generation.
Because, due to overfishing, it is less profitable to fish, many do it illegally; and to boost profits essentially enslave those who work on their boats.	Coastal, marine, and ocean engineering are essential specialties that young engineers need to become proficient in. Activities such as mistreating fisherman can be referred to as "environmental slavery"
Trawling should be banned because it hurts sensitive ecosystems such as coral reefs, destroys habitat for marine life. and releases sediment	Engineers should advocate for elimination of trawling, particularly bottom trawling

Earth Justice, which was founded in 1971 and employs numerous attorneys, is committed to providing legal services towards resolving critical environmental causes, free of charge.

Info from Earth Justice	Written by the author
Saved and restored irreplaceable wetlands and everglades in Florida	Until the 1960s wetlands were referred to as swamps which should be destroyed rather than protected
Earth Justice prevented oil companies from obtaining permits to use seismic air guns. Use of them was a first step towards allowing offshore oil drilling. In addition, the sound of the blast threatened the endangered North Atlantic Right Whale because large marine mammals rely on sound transmission to communicate. Interrupting this process threatened their lives.	Engineers are encouraged to prevent the use of seismic air guns
Earth Justice challenged unfair solar rate hikes in Nevada	Making clean energy less attractive and less cost effective ultimately leads to the increased use of fossil fuels
Earth Justice stopped oil and gas drilling in the Arctic Ocean	These activities pollute water and harm marine life
Earth Justice established the right to ban fracking in New York	Fracking, which is the process of drilling into deep rock formations and inserting pressurized mixtures of water, sand, and chemicals, can contaminate groundwater and induce seismic activity
Earth Justice protected sea turtles in the Gulf of Mexico by suing the National Marine Fisheries Service	This prevented longline fishing which collected more than the allowable number of endangered sea turtles

Earth Justice validated the Civil Rights Act Title VI protections	The Civil Rights Act Title VI of 1964 remains the primary civil rights legislation in the U.S.
Earth Justice protected agricultural worker's rights	Agricultural workers, who are usually minorities, are often exposed to harmful pesticides and grueling working conditions

Chapter 9

Functional Capacity Evaluations

An engineer needs to be physically and mentally fit to perform his job functions. He must be aware of the physical and mental condition of those he has oversight of. In addition, he must also evaluate his own condition and forego opportunities he is not able to safely perform.

Good functional capacity will empower engineers to formulate ethical standards and solutions to address societal changes posed by global warming and war. It will elucidate the contribution of values, morality, and religion towards maintaining a high level of respect for engineers.

These describe situations in which an individual is no longer fit to perform his assigned tasks. An engineer has an ethical obligation to remove an employee from a project location if they have lost the physical, mental, or emotional capabilities needed to travel through factories, laboratories, construction sites, or other hazardous locations in a safe manner. An engineer also must avoid assignments which would require him or her to visit a project location that he is not fit to work on.

Following are a list of criteria and a discussion of how they are applicable.

Lifting ability - office	Those who work in an office are not required to be able to lift heavy objects. They should have the strength to lift laptop computers, small stand-alone printers, partially full file boxes, books, office supplies, and emergency equipment. They are not expected to lift large bottles of water for water fountains. If someone cannot lift these relatively light items that are routinely used in an office environment, they may not be able to work in that location.

Lifting ability - engineer	Engineers are required to visit construction sites, factories, and laboratories on occasion. It is necessary to have the strength to move a small ladder to a location where it is needed. Engineers may be required to move items out of a path of travel when performing an inspection. They need to be strong enough to lift a portable fire extinguisher, which often weighs 35 pounds or more, and assist a person who has collapsed or is injured. Some manual roll up and industrial sliding doors require a significant amount of strength to open and close. If an engineer cannot perform those tasks, he should not accept assignments that require him to work in these locations.
Lifting ability - construction worker	There are several jobs on construction sites that do not require a significant amount of strength. These include inspector, superintendent, flagman, elevator operator, nurse, industrial hygienist, and food service worker. Those performing physical work such as a carpenter, plumber, or electrician require a significant amount of strength. The worker should not be expected to violate OSHA safe lifting guidelines. As an example, a worker may be required to lift a bag of Portland cement which weighs almost 100 pounds. He needs to be able to lift that amount for several hours per day. If a person is unable to do so he cannot be a construction worker.
Lifting ability - factory worker	A factory worker must be able to lift objects of different weights for eight hours per day without suffering fatigue.
Turning, twisting, and bending ability	A person needs to have the flexibility to turn, twist, and bend without sustaining an injury.

Personal finances	It is generally unethical to be financially insecure, live month-to-month, be behind on paying bills, or be insolvent most of the time. It is understood that there are circumstances that can strain a person's finances such as taking care of parents, college expenses, illness, unemployment, underemployment, moving a family member into a nursing home, drastic changes in the job market, divorce, damage to a home, significant car damage, criminal actions, civil actions, severe recession, medical bills, severe drop in the value of real estate, and catastrophic drops in national and international stock market values. However, people should plan for these circumstances so they can recover in a relatively short period of time. Financial problems can impact a person's ability to take care of their physical needs, impair their ability to travel, result in creditors contacting their employer, lower their credit rating, prevent them from having a bank account, cause them to miss office social events, prevent them from obtaining a credit card for business travel, cause physical and mental stress, damage professional relationships, make it difficult to focus and be productive, increase absenteeism, and cause a person to make mistakes. An engineer has the ethical obligation to discuss these matters with someone who is having long term financial problems and attempt to assist them if they are unable to work productively and safely.
Mental disability	Mental disability is a physical condition that affects the mind such as neurological illnesses, traumatic brain injury, low IQ, down syndrome, and genetic disorders. A mental disability can

	affect a person's ability to work productively and safely.
Alcohol abuse	Can affect an employee's ability to work safely and productively
Illegal drug use (Including marijuana)	Can affect an employee's ability to work safely and productively
Prescription drug use	Can affect an employee's ability to work safely and productively
Mental illness/disorder	Can affect an employee's ability to work safely and productively
Permanent or long-term physical injury or condition such as obesity, poor eyesight, and chronic fatigue	Can affect an employee's ability to work safely and productively
Alzheimer's	Can affect an employee's ability to work safely and productively
Dementia	Can affect an employee's ability to work safely and productively
Inability to drive	Can affect an employee's ability to travel to meetings inspections
Family problems	Can affect an employee's ability to work safely and productively
Balance problems	Can affect an employee's ability to work safely and productively
Chronic Pain	Can affect an employee's ability to work safely and productively

Chapter 10

Scenarios

Table #1:

CATEGORY:	Company Certification
TITLE:	Lack of certification
DESCRIPTION:	An engineer was offered an assignment that the company has not certified him to do. He is more than qualified to perform the work. May he accept the assignment?
DISCUSSION:	It is not acceptable for him to accept the assignment. The company has protocols in place for administrative, quality assurance, and financial reasons. The engineer would be taking advantage of the employee who inadvertently offered him the assignment. The appropriate action would be to inform the employee that he is not certified but ask how to become so.
CONCLUSION:	The engineer should not accept the assignment

Table #2:

CATEGORY:	Honesty
TITLE:	Falsifying qualifications
SCENARIO:	A company owner asked an engineer to prepare marketing material which states that the company has designed 26 sewage treatment plants (wastewater treatment plants) when they have only designed 5. The owner tells the engineer that the firm is fully capable of designing the plant but will not be awarded the

	contract if their experience is limited. Can the engineer do so?
DISCUSSION:	Engineers are bound by the requirement to tell the truth even if they believe it will not hurt anyone. People eventually find out that someone has lied. The engineer should ask the owner if he could state they have only designed 5 plants but provide much more detail on the firm's qualifications
CONCLUSION:	The engineer cannot claim they have designed 26 plants

Table #3:

CATEGORY:	Product tolerances
TITLE:	A manufactured item was too thin
SCENARIO:	The specification stated that the thickness of the material could not be any thinner than required. Another engineer noted the discrepancy but felt it was acceptable. A licensed engineer signed the drawings and expected the work to be done exactly as he had designed. It is understood that engineers select dimensions that are consistent with building codes, industry standards, their calculations, and professional judgement. If it was acceptable for an item to be slightly thinner than he designed, he would have stated that in the specifications.
DISCUSSION:	The engineer who noted the discrepancy felt the requirement was arbitrary and the product manufacturing was adequate. The engineer specified the thickness for the item to operate safely
CONCLUSION:	The work needed to be redone

Table #4

CATEGORY:	Homeless Encampment
TITLE:	Displacement of a homeless camp
SCENARIO:	Redesign of a municipal park would cause a large number of homeless people to vacate a park they had been living in for over 10 years. There are no shelters or facilities to assist homeless people in that county
DISCUSSION:	The homeless were camping there illegally and could not claim squatters rights or homestead the property. They routinely left trash and hazardous items such as needles or broken glass on the ground. People were reluctant to visit the park because of their presence
CONCLUSION:	The engineer may redesign the park

Table #5:

CATEGORY:	Conducting Business in a Foreign Country
TITLE:	U.S. based company building in a U.S. territory
SCENARIO:	A U.S. based company with no physical presence on a U.S. territory planned to build a subdivision on the island. The territory licenses engineers. May the U.S. company hire a licensed engineer to design the project and oversee construction.
DISCUSSION:	The engineer is assumed to be competent because he is a licensed engineer. The fact that he is being hired to provide construction management as well as engineering provides a high degree of certainty that the project will succeed. Can the U.S. company proceed with their plans?

CONCLUSION:	The U.S. company may build in the territory

Table #6

CATEGORY:	Civil Rights
TITLE:	A white employee used a racial slur when addressing a minority person
DESCRIPTION:	A construction superintendent heard a white man address a minority person with a serious racial slur.
DISCUSSION:	The superintendent informed the company owner of the white man's conduct
CONCLUSION:	He handled the situation properly

Table #7

CATEGORY:	Community Involvement
TITLE:	Expectation to patronize local businesses
DESCRIPTION:	Several local business owners near the construction site approached the superintendent and said they expected the workers to eat at their restaurants, shop at the clothing store, and purchase goods from their building supply business
DISCUSSION:	It is entirely up to the workers as to where they eat and where they shop
CONCLUSION:	No one is required to patronize local businesses

Table #8

CATEGORY:	Civil Rights

TITLE:	Racist flag at friend's house
DESCRIPTION:	An engineer was invited to a co-worker's house to watch a sporting event. He noticed that he had a racist flag flying from a pole underneath the U. S. flag
DISCUSSION:	An engineer should not associate with racists at work or on their own time
CONCLUSION:	The engineer should leave the property

Table #9

CATEGORY:	Conflict of Interest
TITLE:	Engineer owns property close to project site
DESCRIPTION:	An engineer owns two pieces of commercial property very close to a proposed project site. He informed the owner of this, and the owner replied that it did concern him. The engineer, as always, is committed to remaining objective during the planning, design, and construction phases of the project. Can the engineer accept the project?
DISCUSSION:	The engineer fulfilled his obligation and will remain objective so he can accept the project
CONCLUSION:	He may accept the project

Table #10

CATEGORY:	Green Living
TITLE:	Disagreement with green design expectations
DESCRIPTION:	The city stated on their website that they endeavor to design and build property in a sustainable manner. They did not mention any regulations or specific guidelines.
DISCUSSION:	An engineer decided that to build part of a structure in a green manner would compromise safety.
CONCLUSION:	An engineer doesn't always need to build green

Table #11

CATEGORY:	Advertising
TITLE:	Examination preparation guidebooks
DESCRIPTION:	The author of a Fundamentals of Engineering exam review manual wrote on the cover that we are very experienced and will seriously endeavor to help you pass.
DISCUSSION:	They made no guarantees. They can fulfill what they said they would do
CONCLUSION:	This advertising was ethical

Table #12

CATEGORY:	Community Safety
TITLE:	Damaged tires
DESCRIPTION:	While away from work an engineer noticed a

	parked car with the owner inside that had steel belts showing on two tires. The engineer knocked on the window and told the driver about the condition of the tires. The driver was embarrassed and annoyed but reluctantly thanked the engineer for telling him
DISCUSSION:	The engineer was obligated to tell the owner even if their response could be negative
CONCLUSION:	We must report even if the driver gets angry

Table #13

CATEGORY:	Artificial Intelligence (AI)
TITLE:	Requirement to understand the limitations of AI
DESCRIPTION:	An employee regularly used AI to write reports, schedules, and budgets. He performed only a cursory review of the output. The AI software made an error which caused a four-day delay of critical path work
DISCUSSION:	AI has limited use in engineering and construction. All AI output needs to be checked
CONCLUSION:	We should understand the limitations of AI

Table #14

CATEGORY:	Woman's Rights and Needs
TITLE:	Not Respectful of Woman's Biologic Needs
DESCRIPTION:	A female employee was forced to take time off without pay for menstrual distress
DISCUSSION:	Women should be given a few days off each month with pay to cope with menstrual

	distress. They should not have to ask permission, obtain a signature from their supervisor, or even explain why they will be gone. To avoid embarrassment, awkwardness, and bias they should simply be able to leave a yellow stickie on their desk or a note on their calendar that they will return when they are able to. They should not be expected to check-in with the office, return calls, write emails, or review reports during this time
CONCLUSION:	It is wrong to make her take time off with no pay

Table #15

CATEGORY:	Area of Expertise
TITLE:	Mechanical engineer hired to design HVAC
DESCRIPTION:	A mechanical engineer's expertise was limited to machine design. He had no HVAC design experience. He was hoping to add HVAC design to his portfolio. He was asked to design an HVAC system for a large factory that was temperature controlled.
DISCUSSION:	He should work on several projects with an experienced HVAC engineer before accepting a large job such as this
CONCLUSION:	He should reject the offer

Table #16

CATEGORY:	Fraudulent Contractor
TITLE:	Hiring a subcontractor who is dishonest

DESCRIPTION:	A general contractor needed a specialty subcontractor within two days. He was aware of a licensed subcontractor, but he was convicted of fraud twice in the last six years. He did hire him.
DISCUSSION:	The general contractor's reputation is now tarnished because he is now associated with a convicted criminal.
CONCLUSION:	He must separate himself from the subcontractor

Table #17

CATEGORY:	Social Media
TITLE:	Posting negative reviews on social media
DESCRIPTION:	An engineer was upset about how his tax preparer treated him, billed him, and prepared the return
DISCUSSION:	The engineer wanted to post a negative review about the accountant on social media. It is possible that his clients or co-workers would see the post. Is it ethical for him to post a negative review?
CONCLUSION:	Posting negative reviews will hurt his credibility

Table #18

CATEGORY:	Accepting Overpayment
TITLE:	Overpayment for editing services
DESCRIPTION:	A client asked a civil engineer to fact check 2

	pages of 140 words each. This was for a very simple basic book on carpentry. The client did not understand carpentry, engineering, or writing. He told the engineer he would pay him $2,400.00.
DISCUSSION:	The engineer should inform the client that he was offering an excessive amount of money and that it would only take two hours to review it. He should tell him how much fact checkers and engineers charge per hour.
CONCLUSION:	He can accept two hours of his standard fee

Table #19

CATEGORY:	Conflict of Interest
TITLE:	Accepting a bid from a distant cousin
DESCRIPTION:	A member of a bid review team received a bid from a distant cousin. He met him at a family reunion 11 years ago but has not had any contact with him since.
DISCUSSION:	It is not a conflict of interest because there is no close family relationship or friendship
CONCLUSION:	It is acceptable to review a distant cousin's bid

Table #20:

CATEGORY:	Religious Symbols
TITLE:	An employee wore a religious pin at work
DESCRIPTION:	The pin did not reflect bias or violate anyone's civil rights. One employee complained because he disagreed with those religious views

DISCUSSION:	As long as the employee did not try to impose his views on other people the pin was acceptable
CONCLUSION:	It is acceptable to wear a religious symbol

Table #21

CATEGORY:	Criticism of Public Officials
TITLE:	Employee shows no respect for elected official
DESCRIPTION:	An employee regularly came to work and complained about the conduct and strategies of an elected official. He was a member of an opposing political party
DISCUSSION:	We are expected to be respectful of our elected leadership. The employee has a right to disagree with him but those discussions should not be held at work
CONCLUSION:	It is wrong to criticize an elected official at work

Table #22

CATEGORY:	Respectful Supervision
TITLE:	A manager regularly mistreated his staff
DESCRIPTION:	This supervisor was responsible for a small group of engineers. He regularly criticized his staff when in reality their work was quite good. He was profane and belligerent. He showed no concern for the employees' private lives or professional development.
DISCUSSION:	It is necessary for a manager to show compassion to his staff and care about their

	professional and personal lives. Profanity and intimidation are never acceptable.
CONCLUSION:	The senior employee should formally complain

Table #23

CATEGORY:	Public Information
TITLE:	Engineer asked to brief public on a project
DESCRIPTION:	A manager asked the first engineer to state at a public meeting that four drums of hazardous waste were found at an old dump site when in reality there were 31. The first engineer refused so his manager asked a second engineer to lie, which he did. What should the first engineer do?
DISCUSSION:	The first engineer cannot ignore the discrepancy
CONCLUSION:	He should privately inform his managers superior

Table #24

CATEGORY:	Credit for Work
TITLE:	Engineer did not significantly contribute to a paper
DESCRIPTION:	The first engineer was assigned to work on a project with a second engineer but was distracted by other projects. The second engineer did almost all the work. The second engineer planned to publish the paper with only his name as the author. The first engineer asked for his name to be on the paper and the second engineer refused.

DISCUSSION:	When two names are listed on a paper it implies that both made a significant contribution which was not the case in this scenario
CONCLUSION:	Only the second engineer should be credited

Table #25

CATEGORY:	Career Changes
TITLE:	An engineer wants to switch specialties
DESCRIPTION:	He is tired of one branch of engineering and wants to try another one. He is asked to perform a project even though he has no experience, license, or degree in the new discipline and accepts it because he believes he can retrain while he performs the design work
DISCUSSION:	The disciplines in engineering are very different and a person needs to obtain a sufficient amount of education and training to allow him to work in a professional manner. Was his strategy ethical?
CONCLUSION:	No. He needs to train on the new discipline

Table #26

CATEGORY:	Conflict of Interest
TITLE:	Accept a bid from a previous colleague
DESCRIPTION:	The company owner asked an engineer to review a bid proposal. As it turns out it was from someone, he had worked with
DISCUSSION:	His interaction with that previous colleague was strictly professional and they had not seen

	each other since he left the company
CONCLUSION:	It is acceptable to review a bid from a previous colleague

Table #27

CATEGORY:	Interaction with previous colleagues
TITLE:	Asking colleagues for a job
DESCRIPTION:	An engineer left a position because the manager and co-workers were toxic. He began calling people he had worked with recently and asking for employment
DISCUSSION:	It is certainly acceptable to contact a person that you worked with in the past. It is understandable that you left your job to avoid a toxic environment and does not reflect on your professionalism. It is very possible that his company desperately needs engineers to staff a major project
CONCLUSION:	It is acceptable to ask colleagues for a job

Table #28

CATEGORY:	Continuing Education
TITLE:	Plans for continuing education
DESCRIPTION:	A progressive engineer asks each new employee to outline his plan for continuing education for the next 8 years.
DISCUSSION:	It is an ethical necessity to complete continuing education. Many states require engineers to complete a certain number of PDHs to maintain

	their license. The State(s) may require him to complete classes on ethics, sexual harassment prevention, and new laws, but if not, he still should. Next, he must select classes for his specialties.
CONCLUSION:	His ethical obligation is met

Table #29

CATEGORY:	Qualification
TITLE:	Engineer to design a machine
DESCRIPTION:	An engineer is asked to design a relatively small machine to be used as part of a process line. He has designed numerous machines but that was over six years ago. In that time period a few aspects of machine technology have changed. The owner wants the plans and specifications to be finished in 90 days
DISCUSSION:	The owner has given him ample time to become an expert on the new technology and then design the machine. Engineers need to update their skills on a regular basis because in many specialties the technology changes rapidly. Can he accept the assignment?
CONCLUSION:	The engineer can accept the assignment

Table #30

CATEGORY:	Unemployment
TITLE:	Engineer is offered a minimum wage job
DESCRIPTION:	An engineer finds himself unemployed and cannot find a suitable job. He is told by family

	friends and colleagues that it is easy to get a minimum wage job and he should get one to make at least a little money and demonstrate that he has a work ethic.
DISCUSSION:	This is not reasonable because those jobs are physical, dirty, and often dangerous. It is demoralizing for someone with an engineering degree to serve fast food or clean other people's offices. It would be a better use of time to take classes that will complement his experience and look good on his resume.
CONCLUSION:	Engineers should not work for minimum wage

Table #31

CATEGORY:	Informal Advice
TITLE:	Engineers offered advice off the record
DESCRIPTION:	A friend asked a civil engineer for some advice on a structural engineering issue. The civil engineer explained he was not a structural engineer and the information he would give would only be his best guess and asked him not to tell anybody what he said.
DISCUSSION:	There is no such thing as off - the - record in engineering. People will follow your advice, tell others what you said, and document it. In an attempt to help his friend he gave him his best guess as to how to design it. Should the civil engineer have offered structural engineering advice off-the-record?
CONCLUSION:	He should not give structural engineering advice

Table #32

CATEGORY:	Marketing Leads
TITLE:	Employee takes list of contacts when he leaves
DESCRIPTION:	A sales engineer worked for a company for three years. During that time, he developed several clients and also had access to a list of potential clients
DISCUSSION:	Because he worked for the company and his contacts were developed as part of his responsibilities, he cannot take their information or contact any of them
CONCLUSION:	His action was unethical & a breach of contract

Table #33

CATEGORY:	Site and Laboratory Access
TITLE:	Local citizen
DESCRIPTION:	A resident who lived very close to a construction site was unsure if the dust blowing off of the site and water flowing off of the site were contaminated. He approached the engineer at the gate and asked if he could look around. The engineer said he could and verified that the man was wearing suitable clothing and provided him with a hardhat and vest and escorted him around the site. He directed the operators to shut down their equipment and verified that no one could fall into a trench. The visit took 11 minutes. Should the engineer have let the man onsite?
DISCUSSION:	A concerned citizen has the right to observe

	operations. Should the superintendent have let the man onsite?
CONCLUSION:	The man had a right to be onsite.

Table #34

CATEGORY:	Respect Cultural Practices
TITLE:	Support local traditions
DESCRIPTION:	A factory was being built in a neighborhood of people with strong religious beliefs. To comply with their spiritual leader's direction, all businesses were closed on Sundays, 11 government holidays, 23 religious holidays, and the last 5 days of every month so residents could participate in religious activities.
DISCUSSION:	The construction manager determined that he needed to work 6 - 7 days a week, 8 - 10 hours per day for 4 months to finish on time and avoid liquidated damages. He continued to work but significantly lowered the noise level. When we are working in someone's neighborhood, we are a visitor in their home and should act accordingly. We should support their practices and be interested in their culture
CONCLUSION:	He should not have worked on those days

Table #35

CATEGORY:	Local Residents
TITLE:	Noise complaints
DESCRIPTION:	Neighbors complained about noise coming from

		the site almost every day from very early in the morning to late afternoon
DISCUSSION:		The engineer needed to confine construction activity to normal business hours. He was obligated to install sound barriers before he resumed construction
CONCLUSION:		He had an ethical obligation to reduce noise

Table #36

CATEGORY:	Local Community
TITLE:	Supporting local businesses
DESCRIPTION:	Two local restaurant owners, the manager of the town hardware store, and an owner of a clothing store ask him if he would visit their stores. He felt these establishments were poorly run and chose not to patronize them
DISCUSSION:	The engineer listened politely but did not respond.
CONCLUSION:	He had no obligation to shop at their stores

Table #37

CATEGORY:	Demonstrations
TITLE:	Prevented from crossing a picket line
DESCRIPTION:	There were numerous men standing close together and blocking the door. A few looked at him in a menacing manner. He needed to follow up on three serious injuries that had occurred within the past week
DISCUSSION:	He felt terribly intimidated and unsafe. His

	safety was compromised. He left the site and immediately contacted the union president and the police and arranged for them to clear the men from the entrance and escort him from his car, to the site, back to his car and watch him as he drove away from the area
CONCLUSION:	The engineer's actions were prudent

Table #38

CATEGORY:	Public Demonstrations
TITLE:	Crossing a picket line.
DESCRIPTION:	An engineer arrived at a site to perform an inspection of safety practices, environmental compliance, quality of work, and the progress. There were just a few men on the picket line. He did not appear to be in any danger. There was no activity taking place that would make him feel overly uncomfortable
DISCUSSION:	It is not illegal to cross a picket line. His site visit was critically important
CONCLUSION:	Crossing the picket line was ethical

Table #39

CATEGORY:	Public Demonstrations
TITLE:	Joining a project picket at his place of work
DESCRIPTION:	An engineer joined a picket line established by a construction union
DISCUSSION:	He compromised his role as a safety professional and joined in an activity that could

	endanger workers and the public. Engineers should conduct their business without regard to financial matters. He was not acting in his client's best interest by participating in the demonstration
CONCLUSION:	His conduct was unethical

Table #40

CATEGORY:	Professional Reputation
TITLE:	Visiting a disreputable establishment
DESCRIPTION:	It impairs the engineer, his company, and the engineering if he visits locations where illegal, unethical, and immoral activities take place. An engineer was observed in a retail location such as this and although he did not do anything wrong his name was now associated with improper activity. Even a client made a comment about it.
DISCUSSION:	Most people have an ethical, moral, religious, and conservative stance and are unlikely to do business with someone who does not.
CONCLUSION:	He should not have entered this establishment

Table #41

CATEGORY:	Protecting the Public
TITLE:	Prevent people from approaching excavations
DESCRIPTION:	A contractor was digging a large hole in front of an office building. He is obligated to protect office workers and anyone who would access

	the property from injury or death. The contractor was very busy and only worked on this project one to two days a week. Each day he left the area securely barricaded.
DISCUSSION:	Barricades are moved routinely by other contractors needing access, lawn service personnel, curious onlookers, inspectors, or trespassers. They are often moved due to heavy rains and strong winds. Was the contractor operating in an ethical manner?
CONCLUSION:	No. He must check the barricades frequently

Table #42

CATEGORY:	Hazards
TITLE:	Safety hazards not associated with your position
DESCRIPTION:	If you notice that a car that is being driven by someone that you do not know has steel belts showing on two tires what should you do?
DISCUSSION:	Engineers should be as concerned with safety in the general community as they are in their own laboratory or factory
CONCLUSION:	We are obligated to address any safety issue

Table #43

CATEGORY:	Colleagues
TITLE:	Asking a colleague for a job
DESCRIPTION:	You asked a colleague for a job
DISCUSSION:	People tend to work with many different co-

	workers at different times in their career. Those in a niche profession tend to know other professionals in their area quite well. One purpose of career related associations is to meet potential employers and employees. It is advantageous for an employer to hire someone that they know professionally. In this case it is acceptable to ask for a position in his company. You would not be taking advantage of your colleague by asking him for a job.
CONCLUSION:	Your action was ethical

Table #44

CATEGORY:	Networking
TITLE:	Networking is essentially using your friends
DESCRIPTION:	You chose to switch companies to avoid the many toxic leaders and co-workers in your current position. You asked a friend to vouch for you and arrange an interview with their supervisor?
DISCUSSION:	This puts your friend or family member at risk because if you fail in the job it discredits them. Your friend may have felt obligated to help you even if their supervisor discourages that type of action
CONCLUSION:	It was unethical to ask for an interview

Table #45

CATEGORY:	Business Finances
TITLE:	Borrowing from friends & family to incorporate

DESCRIPTION:	With proper planning you should not have to borrow money from your parents or friends to open a business. Your business could fail and then you would not have the funds to repay them. There are multiple legitimate and fair sources of funding.
DISCUSSION:	Ideally you could start the business using your savings or select a business with little or no capital required. Another option is to continue working full or part time until your enterprise becomes profitable
CONCLUSION:	It is unethical to borrow money from relatives

TABLE #46

CATEGORY:	Career Development
TITLE:	Promote each employee to management
DESCRIPTION:	The owner of an engineering firm believes that each employee can learn to be a manager and advance to a leadership position by mid-career. As an outworking of his philosophy, he requires each young engineer to supervise a group of employees for three to six months. Is this well-meaning company owner acting in an ethical manner?
DISCUSSION:	Different engineers have different temperaments and abilities. Each must be guided into a career path that he will enjoy and be successful in. To do less than that will leave the engineer frustrated and ultimately harm the company. Psychologists and counselors can evaluate an employee's emotional proclivity and recommend employers and positions that will be suitable for them.

CONCLUSION:	It is unethical to require all to be managers

www.ingramcontent.com/pod-product-compliance
Lightning Source LLC
Chambersburg PA
CBHW052111070526

44584CB00017B/2434